PLANTS AND FLOWERS

Written by: **Brian Holley**
Supervisor, Children's Garden
Royal Botanical Gardens

Editors: **Nadia Pelowich**
Curtis Rush

Illustrations: **Martin Magee**

Design: **J.T. Winik**
Rick Rowden

Copyright © 1986 by Hayes Publishing Ltd. All rights reserved. No part of this work may be reproduced or transmitted in any form or by any means, electronic or mechanical, including photocopying and recording, or by any information storage or retrieval system, without permission in writing from the publisher.

© 1986 Hayes Publishing Ltd.
Burlington, Ontario

Hayes Publishing Ltd.,
3312 Mainway, Burlington, Ontario L7M 1A7, Canada

Look for this trivia worm throughout this book.

WITHOUT PLANTS WE COULDN'T EXIST!

Today in our cities plants are especially important. They help to filter out the pollution from air and water.

Take a look around you! How many things can you find that are made from plants? Don't forget the paint on the walls, this book or the laces in your shoes.

Oxygen

Fuel

Food

Clothing

This rubber ball could be made from the sap of a rubber tree, but today it is more likely to be made from coal or oil. Both coal and oil are formed from ancient plants.

TWO PLANTS CAN BE AS DIFFERENT AS A CATERPILLAR AND A DOG

It is easy to see the differences between these animals. The differences between plants can be just as great! The mushroom is a fungus. It doesn't produce flowers, seeds or leaves like the sunflower. The biggest difference is in the color. The sunflower has green leaves. The green color is caused by chlorophyll.

3. The leaf releases leftover oxygen into the air for animals to breathe.

The Fly Ageric mushroom is one of many poisonous mushrooms.

PLANTS EAT SUNSHINE!

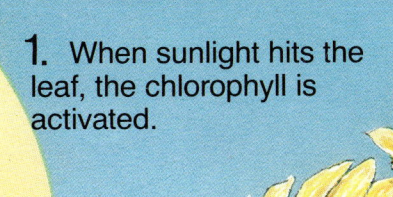

1. When sunlight hits the leaf, the chlorophyll is activated.

Long before the settlers arrived, Indians were making sugar from the sap of trees, like the sugar maple.

2. It breaks water into hydrogen and oxygen, adds carbon from carbon dioxide, and presto, it makes sugar! This is called photosynthesis.

4. Plants eat sunlight energy and we eat plants, so we must be full of sunshine.

5. Sometimes we taste sugar in fruits like apples and oranges, but most of our sugar comes from the stem of a grass called sugar cane and the root of the sugar beet.

SUGAR CANE STEM

FOOD STORAGE

Many of the vegetables we eat are actually plant storehouses.

Kohlrabi is a strange-looking vegetable. Can you guess what part of the plant the food is stored in? (See page 32 for the answer.)▼

Carrots, turnips and beets store food in their roots.
▼

Potatoes are really underground stems called tubers. Potatoes don't taste sweet because plants often ▲ change sugar into starch.

Onions are actually tiny stems covered in scalelike leaves. This type of structure is called a bulb.

6

FROM FLOWERS TO FRUIT

1. For seeds to be produced, pollen from the same type of flower must land on the stigma.

2. Once in place, the pollen grain grows down through the stigma and joins with the ovule.

CHERRY BLOSSOM

ANTHER
OVARY
POLLEN

STIGMA
OVULES
PETALS

Some plants, like the cucumber, have separate male and female flowers. The cucumber plant usually has more male flowers than it needs. Some gardeners pick them off and put them into salads. The flowers are edible and will certainly brighten up your dinner.

Which of these cucumber flowers is female? See page **32**

Try touching the anther of a flower. Often the pollen will stay on your finger. Then, gently dust the stigma with the pollen. The sticky surface of the stigma will hold the pollen in place. Now you are a pollinator. See page 10 for some other pollinators.

3. The seed then starts to develop. As it does, the ovary grows into fruit.

A tasty tea can be made from the flowers of the basswood tree. You may find it in the grocery store, labeled as linden tea.

FLOWERS IN YOUR GARDEN

Flowers are designed to attract pollinators.

Apple blossom

Irises have markings to show bees where to land. Can you see any other flowers with landing strips?

Red and orange flowers are usually pollinated by hummingbirds and butterflies.

Pansy

Iris

Black-eyed Susan

The dark center of the black-eyed Susan provides a target for bees. Bees see the colors yellow and blue, but they can't see red. Try an experiment with bee color vision. Mix two clear glass bowls of sugar and water. Put one on yellow paper and the other on red and leave them outside. Which one does the bee visit?

DESERT PLANTS

Imagine living in a desert. How would you survive without water in a hot, windy environment? To find water, you could try digging a well, or store the scant rainfall. You would likely build a low, stable shelter to break the wind and provide shade. Desert plants use these same strategies to live in this climate.

Photosynthesis, the making of sugar, takes place in the green stems of cacti. The thick outer skin helps reduce water loss, and the short squat stem is very stable during windstorms.

OPUNTIA

The shaggy hairs of "old man" cactus protect it from both wind and sun.

OLD MAN CACTUS

COBIVIA (GOLDEN BARREL CACTUS)

PRICKLY PEAR

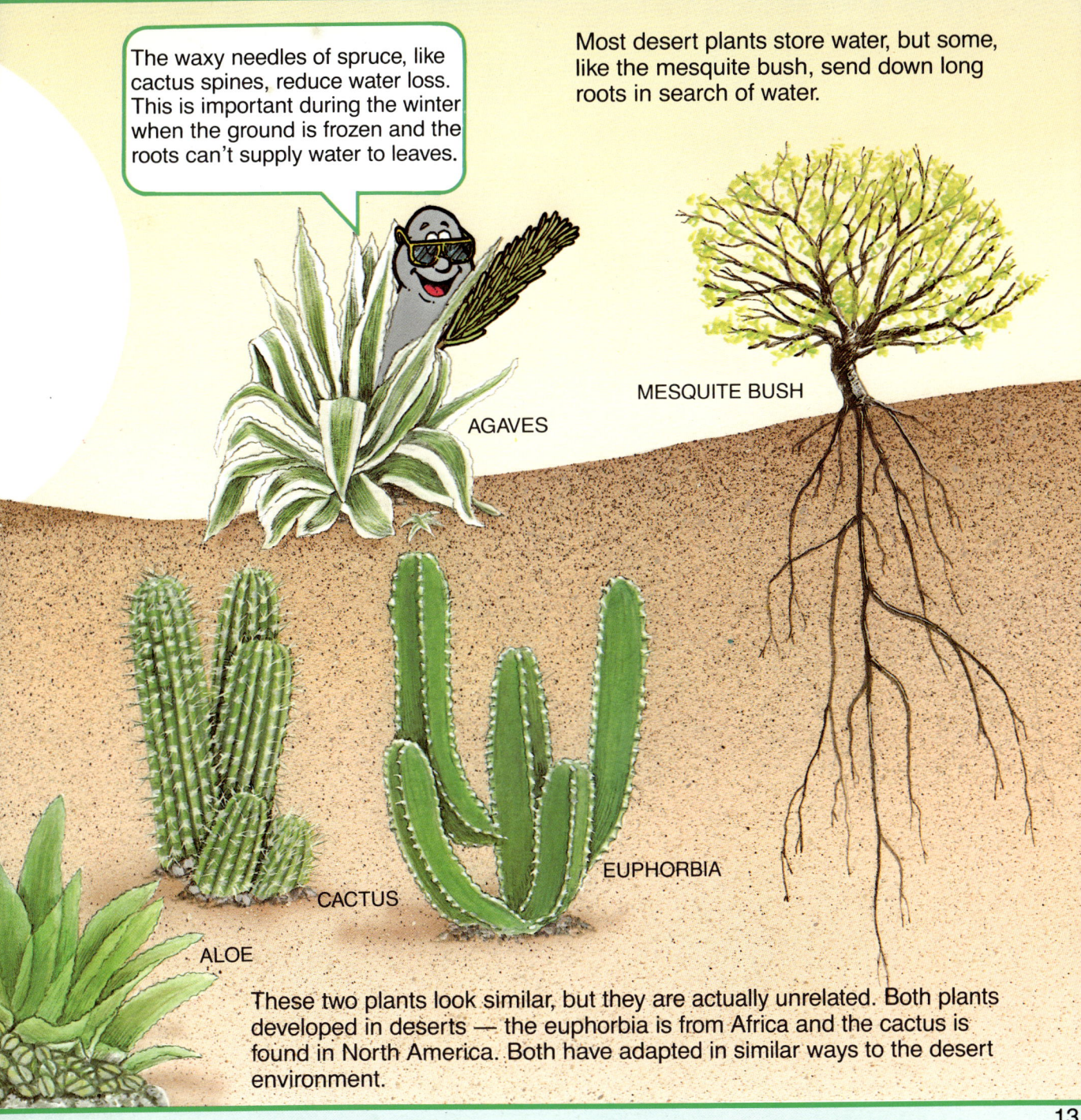

ON THE FOREST FLOOR

The environment on the forest floor is the opposite of the desert. Here it is shady, moist and sheltered from wind. How are the plants of these two environments different?

Why do you think the Indian pipe is so pale? You're right! it doesn't have chlorophyll. Therefore, it can't eat sunshine. Instead, the Indian pipe lives on dead leaves. ▼

The large, dark green leaves of the Christmas fern make the most of limited sunlight. ▲

Most forest wild flowers are produced early in the spring. Then before the trees have leaves, it is bright in the forest and most plants take advantage of the sun to produce flowers. In the summertime, the trees have fully grown leaves and the forest is shady. There isn't enough sunlight coming through the leaves for most plants to grow.

"Many of our forest trees produce edible nuts. Beechnuts, walnuts, butternuts and hickory nuts all taste good, but you have to beat the squirrels to them!"

Forest soils are good examples of recycling. A whole world of creatures busily turns forest litter into soil. When you look closely among the leaves, you will see some of these soil makers.

SOIL MAKERS

The thin, white net on this leaf is part of a mushroom. The mushroom is breaking down the dead leaf into food for itself.

▼ This is a millipede. Its name means a thousand feet. How many feet does it really have?

Springtails are the most abundant insects in the world. There could be thousands of them in your backyard. ▲

◄ Some sow bugs curl up into armored balls when danger approaches.

The soil produced by soil makers is called humus. It is full of the nutrients plants need for healthy growth. Just as we need vitamins to grow, plants need good soil to grow.

15

THE SOIL CHEF

You could make your own soil by recycling kitchen scraps!

Recipe:
Collect two cups of kitchen scraps. (Eggshells, coffee grounds, fruit and vegetables will work well.) Tear into small pieces and place in a plastic bag.

Add 250 mL (1 cup) of garden soil and enough water to make the mixture moist but not wet. Blow a little air into the bag, seal with a twist tie and give it a good shake.

Shake the bag once a day, and every week open it to allow some fresh air in.

In a few weeks, the microscopic soil makers will have turned the mixture into soil. This type of soil is called compost.

Most northern tree seeds won't grow until they have been kept cold for at least a month.

Can you guess why? See page 32.

GROWING WITH GARBAGE

NOW THAT YOU HAVE SOME SOIL, TRY SOME MORE RECYCLING.

Try planting grapefruit or orange seeds! Margarine tubs, paper cups and many other disposable containers make good pots. Poke holes in the bottom for drainage, fill with soil and bury the seed 1 cm deep. Place the container in a plastic bag until the seed starts to grow.

Carrot tops make an attractive plant. Cut 3 cm (1 ¼ in.) off the top of a carrot. Eat the bottom and plant the rest in the soil so that just the very top is showing. Soon, you will have an attractive, new plant!

Even if you don't have soil, you can start an orange seed in a used tea bag. Keep the bag in a jar so it won't dry out.

Don't throw out old bleach bottles because they can make beautiful planters like this one. You can cut the bleach bottle in different ways.

Dish-soap bottles are terrific watering cans, and window spray bottles are good misters. Important: Wash them out well before using them.

17

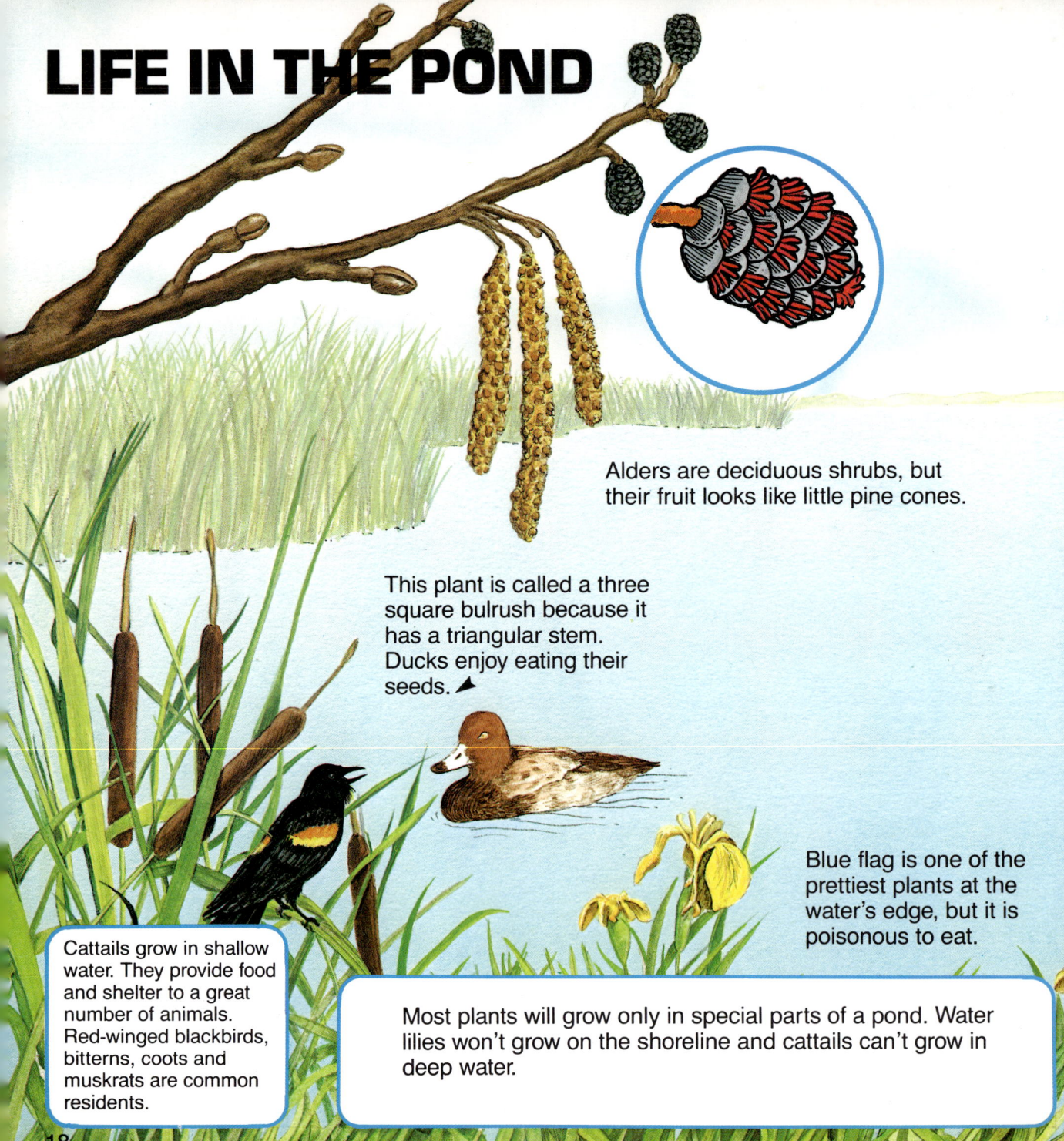

The knees of the bald cypress are actually roots that grow out of the water to obtain air.

The yellow bladderwort has small, round traps on its leaves. These capture tiny water creatures, providing the plant with necessary nutrients. When an animal touches the trigger hair, the door slams shut, trapping it inside.

These greenish clouds are made up of microscopic plants called algae.

MICROSCOPIC VIEW OF DUCKWEED FLOWER

Water lilies have tubes in their stems to channel air to their submerged roots.

Tiny duckweed plants form thick mats on the surface of the pond. One type, Wolffia, is the smallest flowering plant in the world. You would need a microscope to see its flowers!

In submerged plants like the coontail, photosynthesis takes place under water.

19

The kangaroo vine will quickly climb up a trellis.

Jade plants have fat stems and leaves. Would they grow in the forest or desert?

Spider plants produce young plants from runners. They grow best in hanging baskets.

The piggyback plant grows wild along the west coast of North America. Can you guess how it got its name?

MAKING NEW PLANTS

Vegetative propagation uses a plant part other than the seed to produce a new plant. Plants identical to the plant they were taken from are produced. These new plants are called clones.

African violets, jade plants and panda plants are often produced from leaf cuttings.

The young plantlets of the spider plant and the piggyback plant are easily propagated.

Using a pair of scissors, cut 7.5 cm (3 in.) off the tip of a stem and remove the bottom leaves.

If you cut yourself, would you put garden soil on it? Not likely because the cut would become infected. If you use garden soil for your cuttings, they would become infected as well. Sand, perlite, peat moss and vermiculite all act like a bandage for our cuttings, protecting them from infection.

Stick the stem of a leaf cutting into moist sand or vermiculite so that only the leaf shows above the surface. It will take root in about six weeks and can then be replanted in earth. Many plants are propogated by stem cutting, as well.

Small propagation units like this one will help keep the air around the plant moist. This is important because the cuttings don't have any roots to provide water to the leaves.

TRIVIA
Black spruce trees have drooping branches. Sometimes, the bottom branches get buried and new trees start growing from them.

Plant the cutting up to the leaves.

Put a plastic bag over the pot and cutting. Blow some air into it and seal the bag with a twist tie.

Cuttings can also be started in water. Fill a bottle with water and set the bottom of the cutting in it. Do roots grow faster in water? Are the roots different?

WEIRD CACTI

Another form of vegetative propagation is called grafting. To start, we need two plants that are closely related. Then, we take a branch from one plant and notch it into the other plant.

Cacti are members of a group of plants known as succulents. Succulents means "juicy plants."

Take two cacti; they should be similar in size and shape, but one could be fuzzy and the other prickly.

With a clean, sharp knife, make a flat cut through the top of each cactus. In the center, you will see a core.

Exchange tops, lining up the core as closely as possible. If the two cut surfaces don't line up closely, trim the edge a little at a time until they do. Then, place your cacti creatures in a corner, out of bright light, where they won't be disturbed. They should join in about two weeks.

MINI GREENHOUSES

PICKLE JAR TERRARIUM

You will need: 1 big glass jar
                 pebbles
                 potting soil
                 small plants or rooted cuttings
     Optional: small pieces of rock,
                 makeup mirror, tiny
                 figurines

(1) Place the jar on end. Set the pebbles in the bottom to a depth of 2 cm (¾ in.).

(2) Put 6 cm (2¼ in.) of soil in the jar. Then, place the plants in to make a miniature landscape. A small mirror will look like a pond, and figurines or rocks will add interest to the scene.

(3) Use an old dish-soap container to mist the terrarium lightly and put the top on. Keep your terrarium away from direct light or the plants will get too hot.

PLANT HOSPITAL

This plastic bag greenhouse uses two coat hangers for its frame. It is a great place for sick plants. It is warm inside, and there is lots of moisture in the air. Keep your plants inside when you go on holidays, too! Ask your parents to help you with this.

SPROUTS

You can garden year round with sprouts! Most health food or bulk food stores will have a wide variety of seeds. Try wheat, beans, lentils or chickpeas; they will all work.

Day 1—Soak the seeds in warm water overnight.

Day 2—Put the seeds in a strainer and rinse with cold water. Cover with a damp cloth.

WARNING: Don't use garden seeds! (They may be sprayed with poison!)

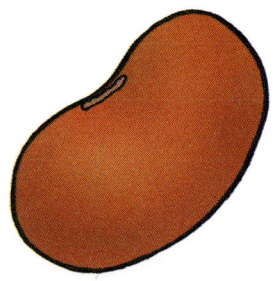

(1) The hard outer covering of the seed is called the seed coat.

Repeat this process each morning and evening. In a few days, they will be ready to go in a salad, on a sandwich or in a cooked omelet.

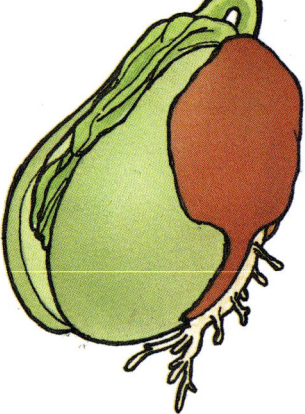

(3) A tiny root is poking out. It's alive! You could eat them at this stage, but let's see what happens.

HAPPENS

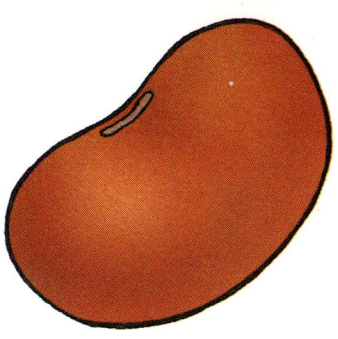

(2) The water swells the seed, softening the seed coat.

(4) The stem is bending up. The thick storage pouches are still attached to the young seedling, providing it with food it needs to grow. Better eat them now before they get too big.

TRANSPLANTING BEANS

Fill a grocery bag half-full of soil and seal with a twist tie. Put near a window or outside in the warm weather. Place a sheet of plastic underneath to catch drips.

Lay the bag on its side and make a couple of short slits along the bottom on both sides. These provide drainage.

Make three holes in the top and transplant your seedlings into them.

Plant them so the roots are completely covered and the leaves are above the soil.

Watch them grow!

Beans on trees? Both black and honey locust trees are in the bean family, but don't eat the seeds! They're poisonous!

SNEAKY SEEDS

How many seeds are in this apple?

How many apples are on this tree? Guess that there are 100. If that were true, this tree will produce 500 seeds.

Young seedlings are often eaten by rabbits.▼

What would happen if all these seeds started to grow?

Overpopulation! The seedlings are so crowded that they could never grow into trees. What's the solution?

Many seeds won't land in places where they can grow, and even if the seedlings have a chance to grow into small plants, they may be attacked by insects or disease.▼

People often use the same name for two different plants.

Black-eyed Susan could be either a cone flower or a thunbergia.

Ironwood is a name used for several trees with very hard wood.

In order to reduce confusion, botanists have special names for each plant. They are called scientific names and are made up of two words. The FIRST is the group that the plant belongs to and the SECOND is a specific type of plant in that group. The names are in Latin, an ancient language. Latin is understood by scientists of all nationalities.

PLANT SCIENTIFIC NAME AND EXPLANATION

Yarrow — *Achillea millefolium.* Achillea is a group name. Yarrow is said to have been used by the Greek hero, Achilles, to cure the wounds of his soldiers. *Millefolium* means thousand leaves. So, the name tells us a little about this plant.

Acer saccharum: sugar maple. The name tells us that these trees have hard wood and produce sugar. Acer rubrum means red maple. The word rubrum means red-flowered. Red maples have beautiful red flowers in early spring.

Mums — *Chrysanthemum* means gold-flowered and *leucanthemum* means white-flowered. That's a pretty good description of the gold and white flowers of the daisy, or as scientists would say, *chrysanthemum leucanthemum.*

Rosemary: *Rosmarinum officinalis.* Rosmarinum means sea spray. It grows wild along sea cliffs. Officinalis tells us it was once used as an official medicine.

Touch-me-not — Impatients is the group name. They really aren't very patient. All you have to do is touch the seed pod and it explodes. Remember the apple? Why do you think the touch-me-not would do that?

Sometimes, scientific names tell us where the plant grows. Where do you think the goldenrod, *Solidago canadensis,* would be common? *Solidago* means healing plant. Perhaps it was used as a medicine by the pioneers.

Sometimes the scientific names tell us what the plants smell like. *Symplocarpus foetidus* is skunk cabbage. *Foetidus* means stinky, and it is true, the skunk cabbage does smell. *Symplacarpus* is a mouthful, but it just means that the fruits are all joined together.

GLOSSARY

ADAPTATIONS: Special features of a plant or animal that improve its ability to survive. Example — flowers that use markings to attract pollinators.

BUD: An undeveloped plant shoot that contains a leaf or flower.

CHLOROPHYLL: The green-colored matter that enables plants to convert sunlight, water and air into sugar.

CONIFEROUS: Trees and shrubs in which the seeds develop in cones. Most conifers are evergreen trees.

DECIDUOUS: Trees and shrubs that shed their leaves each year.

ENVIRONMENT: Everything that surrounds a plant or animal. Climate, soils and other plants and animals are all part of a particular environment.

GRAFTING: The joining of a part of one plant to another plant.

LEAF: Usually a thin, flat, green structure where photosynthesis takes place.

OVULE: The part of a flower that develops into a seed.

PETIOLE: A leaf stalk.

PHOTO-SYNTHESIS: This is a process that happens in green plants or in plants that have chlorophyll. It is used to produce sugar from light, hydrogen from water and carbon from carbon dioxide in the air.

POLLEN: A fine powder produced by a flower that is carried to other flowers by pollinators. A pollen grain joins with an ovule to form a seed.

PROPAGATION: The production of living things.

ROOT: The part of a plant that grows underground, providing the plant with water and nutrients from the soil, as well as support.

STEM: The part of a plant from which branches, leaves and flowers grow.

TUBER: A short, fat stem (usually underground) that has tiny buds on its surface.

ANSWERS

1. Food is stored in the short, ball-shaped stem of the kohlrabi.
2. The female cucumber flower has a tiny cucumber growing from it.
3. Most seeds develop in late summer and fall. What would happen to a tiny seedling if it started to grow in the autumn? It would be damaged by the winter weather. Seeds that require cold weather to germinate can't start growing until spring, when warm temperatures and lots of rain help them get a good start.